BIOTECHNOLOGY IN AGRICULTURE,
INDUSTRY AND MEDICINE

FLAX LIPIDS: CLASSES, BIOSYNTHESIS, GENETICS AND THE PROMISE OF APPLIED GENOMICS FOR UNDERSTANDING AND ALTERING OF FATTY ACIDS

BIOTECHNOLOGY IN AGRICULTURE, INDUSTRY AND MEDICINE

Additional books in this series can be found on Nova's website under the Series tab.

Additional E-books in this series can be found on Nova's website under the E-book tab.

BIOTECHNOLOGY IN AGRICULTURE,
INDUSTRY AND MEDICINE

FLAX LIPIDS: CLASSES, BIOSYNTHESIS, GENETICS AND THE PROMISE OF APPLIED GENOMICS FOR UNDERSTANDING AND ALTERING OF FATTY ACIDS

BOURLAYE FOFANA
SYLVIE CLOUTIER
AND
RAJA RAGUPATHY

Nova Science Publishers, Inc.
New York

Copyright © 2011 by Nova Science Publishers, Inc.

All rights reserved. No part of this book may be reproduced, stored in a retrieval system or transmitted in any form or by any means: electronic, electrostatic, magnetic, tape, mechanical photocopying, recording or otherwise without the written permission of the Publisher.

For permission to use material from this book please contact us:
Telephone 631-231-7269; Fax 631-231-8175
Web Site: http://www.novapublishers.com

NOTICE TO THE READER

The Publisher has taken reasonable care in the preparation of this book, but makes no expressed or implied warranty of any kind and assumes no responsibility for any errors or omissions. No liability is assumed for incidental or consequential damages in connection with or arising out of information contained in this book. The Publisher shall not be liable for any special, consequential, or exemplary damages resulting, in whole or in part, from the readers' use of, or reliance upon, this material. Any parts of this book based on government reports are so indicated and copyright is claimed for those parts to the extent applicable to compilations of such works.

Independent verification should be sought for any data, advice or recommendations contained in this book. In addition, no responsibility is assumed by the publisher for any injury and/or damage to persons or property arising from any methods, products, instructions, ideas or otherwise contained in this publication.

This publication is designed to provide accurate and authoritative information with regard to the subject matter covered herein. It is sold with the clear understanding that the Publisher is not engaged in rendering legal or any other professional services. If legal or any other expert assistance is required, the services of a competent person should be sought. FROM A DECLARATION OF PARTICIPANTS JOINTLY ADOPTED BY A COMMITTEE OF THE AMERICAN BAR ASSOCIATION AND A COMMITTEE OF PUBLISHERS.

Additional color graphics may be available in the e-book version of this book.

LIBRARY OF CONGRESS CATALOGING-IN-PUBLICATION DATA

Fofana, Bourlaye.
  Flax lipids : classes, biosynthesis, genetics and the promise of applied genomics for understanding and altering of fatty acids / Bourlaye Fofana, Sylvie Cloutier and Raja Ragupathy.
       p. cm.
  Includes index.
  ISBN 978-1-61761-470-5 (softcover)
  1. Fflax. 2. Linseed oil. I. Cloutier, Sylvie. II. Ragupathy, Raja. III. Title.
  QK495.L74F64 2010
  633.8'5--dc22
                                                                                           2010031188

Published by Nova Science Publishers, Inc. † New York

CONTENTS

Preface		vii
Introduction		1
Chapter 1	Lipid Classes, Localization, Biosynthesis and Regulation	3
Chapter 2	Current Trends and Future Perspectives in Applied Genomics and Transgenic Research in Flax	17
Conclusion		29
References		31
Index		45

PREFACE

Flax (*Linum usitatissimum* L.) is a multi-purpose crop with emerging markets for its industrial, food and feed uses. Linseed oil extracted from seeds is used for environment friendly industrial applications such as linoleum flooring, manufacturing of paints, stains and varnishes. Flax seeds are also popular in food products and as nutraceuticals. Health promoting virtues of flaxseeds have been attributed to their essential fatty acids (EFAs) comprising omega-6 (linoleic; C18:2) and omega-3 (α-linolenic; ALA; C18:3) fatty acids (FAs), ample dietary fibres and cancer-fighting lignans among others. Flax EFAs are important players in many metabolic processes in cell biology not only as cell membrane structural components but also as storage lipids. FA biosynthesis in oilseeds, including flax, is catalyzed by a set of condensing and desaturation enzymes located in plastids and the endoplasmic reticulum. The free FAs are involved in several metabolic pathways such as storage lipid synthesis in developing seeds, cutin and suberin production, and cuticular wax layer synthesis in different tissue types and developmental stages. A thorough understanding of these multiple aspects of flax lipids is paramount to the development of strategies to modify such essential metabolic components. This review summarizes current knowledge of different lipid classes found in flax, their localization and genetic control, and the regulation of flax seed lipid biosynthesis (in planta). Use of genomics for understanding the genetics of flax FAs and its potential for alteration of the FA profiles towards meeting specific end uses is discussed.

INTRODUCTION

Cultivated flax (*Linum usitatissimum* ssp *usitatissimum* L.) is an annual self pollinated crop belonging to the *Linaceae* family. Flax has been grown either for its stem from which fibres are extracted (fibre flax, [*Linum usitatissimum* L. convar. elongatum (Vav. & Ell.)]) or its oil producing seeds (linseed or oilseed flax, [convar. mediterraneum (Vav. & Ell.) Kulpa & Danert]) for more than a thousand years (Zohary 1999). The plant origin and history, genetics and breeding, biological description and economic importance have been recently reviewed (Cullis 2007). While not disregarding the importance of fibre flax, this review will focus specifically on oilseed flax (also known as linseed) with regards to its lipid composition, its major lipid classes, the localization and biosynthesis of FAs having major economic values and, current and future trends in applied flax genomics.

Chapter 1

LIPID CLASSES, LOCALIZATION, BIOSYNTHESIS AND REGULATION

Lipids are FAs and their derivatives, as well as substances that are related biosynthetically or functionally to these compounds (http://www.lipidlibrary.co.uk/Lipids/whatlip/index.htm) and are important forms of carbon storage in plant seeds and cellular membranes (Thole and Nielsen 2008). Broadly, lipids are readily soluble in organic solvents such as chloroform, benzene, ethers and alcohols. FAs are natural compounds synthesized by condensation of malonyl CoA units by a FA synthase complex. They usually contain even numbers of carbon atoms in straight chains (commonly C4–C24) that may be saturated or unsaturated, and may contain a variety of substituent groups (Voelker and Kinney 2001). Lipids contribute to vital physiological functions such as seed germination, desiccation (Baur 1998, Slabas et al. 2001, Buschhaus et al. 2007), cell, tissue and organ structural integrity (Tang et al. 2005) and protection against biotic and abiotic stresses (Chapman et al. 1998, Metz et al. 2000, Van der Selt et al. 2000, Chechetkin et al. 2008, Pollard et al. 2008, Chechetkin et al. 2009). As in many other plant species, flax lipids are found in different parts of the plant in the forms of free FAs, triacylglycerols (TAGs), oxylipins, waxes and many other lipophilic compounds.

1.1. FLAX SEED LIPID CLASSES

With the dawn of the lipidomics era, a growing number of lipids are expected to be discovered and massive amounts of data generated. A new comprehensive classification and nomenclature system, Lipids Metabolites and Pathways Strategy (LIPIDS MAPS), was developed in 2005 (Fahy et al. 2005). For classification purposes, lipids have been finely defined as hydrophobic or amphipathic small molecules that may originate entirely or in part by carbanion-based condensations of thioesters (fatty acyls, glycerolipids, glycerophospholipids, sphingolipids, saccharolipids, and polyketides) and/or by carbocation-based condensations of isoprene units (prenol lipids and sterol lipids). Moreover, lipids have been broadly subdivided into "simple" and "complex" groups, with simple lipids being those yielding at most two types of products on hydrolysis (e.g., FAs, sterols and acylglycerols) and complex lipids (e.g., glycerophospholipids and glycosphingolipids) yielding three or more products on hydrolysis (Wanasundara et al. 1999, Fahy et al. 2005). The LIPIDS MAPS classification system organizes lipids into eight well-defined categories: fatty acyls, glycerolipids, glycerophospholipids, sphingolipids, sterol lipids, prenol lipids, saccharolipids, and polyketides (Fahy et al. 2005, 2009). Seed storage lipids are the most attractive targets for metabolic engineering because they are less likely to disturb the physiology of the plant, and because they represent the major lipids used in food and non-food applications. This review will focus on fatty acyl and glycerolipid classes (Table 1) which are the most prevalent lipid forms in flax seeds (Wanasundara et al. 1999).

Table 1. Lipid classes and subclasses based on LIPIDS MAPS system[a]

Category	Class[b]	Subclass[c]
Fatty acyls (12)	FA and conjugates (17)	Straight-chain FA
		Methy-branched FA
		Unsaturated FA
		Hydroperoxy FA
		Hydroxy FA
		Oxo FA
		Epoxy FA
		Methoxy FA
		Amino FA
		Carbocyclic FA
		Heteroclyclic FA

Category	Class[b]	Subclass[c]
	Octadecanoids (2)	Dicarboxylic acids
		12-oxophytodienoic acid metabolites
		Jasmonic acids
	Eicosanoids (12)	Prostaglandins
		Thromboxanes
	Docosanoids (1)	Docosanoids
	Fatty alcohols (1)	Fatty alcohols
	Fatty aldehydes (1)	Fatty aldehydes
	Fatty esters (8)	Wax monoesters
		Wax diesters
		Cyano esters
		Lactones
		Fatty acyl CoA
		Fatty acyl-ACP
		Fatty acyl carnitines
		Fatty acyl adenylates
	Fatty amides (4)	Primary amines
		N-acyl amides
		Fatty acyl homoserine lactones
		N-acyl ethanolamides (endocanabinoids)
	Fatty nitriles (1)	Fatty nitriles
	Fatty ethers (1)	Fatty ethers
	Hydrocarbons (1)	Hydrocarbons
	Hydroxygenated hydrocarbons (1)	Hydroxygenated hydrocarbons
Glycerolipids (3)	Monoradylglycerols (5)	Monoacylglycerols
		Monoalkylglycerols
		Mono- (1 Z-alkenyl)-glycerols
		Monoacylglycerolglycosides
		Monoalkylglycerolglycosides
	Diradylglycerols (9)	Diacylglycerols
		Alkylacylglycerols
		Dialkylglycerols

Table 1. (Continued).

Category	Class[b]	Subclass[c]
		1 Z-alkenylacylglycerols
		Diacylglycerolglycosides
		Alkylacylglycerol glycosides
		Dialkylglycerolglycosides
		Di-glycerol tetraethers
		Di-glycerol tetraether glycans
	Triradylglycerols (5)	Triacylglycerols
		Alkyldiacylglycerols
		Dialkylmonoacylglycerols
		1 Z-alkenyldiacylglycerols
		Estolides

[a] Adapted from (Fahy et al. 2005);
[b] Number in bracket indicates the number of classes and subclasses in this category or class;
[c] Only the subclasses most relevant to plants are listed.

1.1.1. Fatty Acyls

Fatty acyls represent the major lipid building block of complex lipids and constitute one of the most fundamental categories of biological lipids (Fahy et al. 2005). In this category are listed FA subclasses including straight-chain saturated and unsaturated FAs, hydroperoxy FAs, hydroxy FAs and epoxy FAs, all representing biologically active lipid compounds either for food or industrial purposes. Most, if not all, of these FAs have been reported in different parts of flax plants (Baertschi et al. 1988, Wanasundara et al. 1999, Morrison and Akin 2001, Guttierrez and Del Rio 2003, Siemens and Daun 2005, Fofana et al. 2006, El-Beltagi et al. 2007, Sharmin et al. 2007). Jasmonic acid and its precursors, 12-oxophytodienoic acid metabolites, two members of the octadecanoid subclass, have been widely reported in flax (Baertschi et al. 1988, Chechetkin et al. 2001, 2008, 2009) where, as oxylipins, they play major roles in several plant developmental processes.

1.1.2. Glycerolipids

Seed storage lipids are found mainly in the form of glycerolipids (Slabas et al. 2001) and the pioneer study by Brockerhoff and Yurkowski (1966) identified mono-, di- and triacylglycerols in flax seed oil. The TAGs accounted by far for the highest proportion of neutral lipids, representing up to 30% of fresh seed weight while mono- and diacylglycerols and free FAs accounted for only 0.34, 0.64 and 0.42 %, respectively (Wanasundara et al. 1999). Recently, using Maldi-RTOF-MS and ESI-IT-MS, more than 20 TAGs have been identified directly from whole linseed oil samples, mainly composed of C18:3, C18:2, C18:1 (oleic acid) moieties, and to a lesser degree of C18:0 (stearic acid) and C16:0 (palmitic acid) moieties (Krist et al. 2006). However, it is worth noting that the FA composition of TAGs is strongly determined by genotypes and environment (see section 2.4).

1.2. LIPID LOCALIZATION

1.2.1. Stem

The flax plant has one main stem (Cullis 2007) and fibre flax is one of the oldest stem fibres known to man (Nandy and Rowland 2008). The main stem of fibre flax includes few branches and reaches approximately 80-120 cm in height. The oilseed flax plant architecture is characterized by more branching and a shorter stature (40-60 cm). The average stem diameter of flax plants is about 2-5 mm. The stem consists of six consecutive layers: epidermis, cortex, phloem, cambium, xylem and pith. Fibre cells originate from the procambial cells in the protophloem which are long, multinucleate cells without septum or partition. It has also been shown that high fibre flax contains more stem area compared to oilseed flax (Nandy and Rowland 2008). By recovering the bast region of fibre and oilseed flax plants, Morrison and Akin (2001) compared the chemical composition of the outer layer (made of the epidermis covered with the cuticle and cortical parenchyma cells) and the fibre bundles. The cuticle is composed of cutin, which is a lipid polymer (Pollard et al. 2008) and a layer of waxy material. The wax extracts from diverse plant species consist of homologous series of very-long-chain FAs, aldehydes, primary and secondary alcohols, ketones, and alkanes of chain lengths C20–C36, as well as C38–C70 alkyl esters (Jetter et al. 2006). More details on cuticular cutin and wax deposition in plant leaves can be found elsewhere (Jetter et al. 2006; Pollard et al. 2008;

Van Maarseveen et al. 2009). In flax stem, wax and cutin related-compounds accounted for the highest proportion of chemical constituents found in the outer layer compared to fibre bundles. Moreover, fibre flax stems contain more wax and cutin related-compounds compared with oilseed flax stems. The total wax related-compounds were 63.6 % in fibre flax variety Laura compared to 36.4% in oilseed flax variety Omega with the corresponding level of cutin related components being 84.1% and 35.1%, respectively (Morrison and Akin 2001). The major components of the outer layer were dihydroxy FAs, accounting for 98-99 % of the FAs. The principal dihydroxy FAs associated with cutin are mixtures of 8,16- and 9,16-dihydroxyhexadecanoic, 8,17- and 9,17-dihydroxyheptadecanoic acids, and 9,18- and 10,18-dihydroxyoctadecanoic acids with the 8,16- and 9,16-dihydroxyhexadecanoic acids being the major components. The principal FAs associated with waxes are mixtures of C12–C28 FAs, with C16 FA being the major component (16.9%) although C28 alcohol represents the most abundant component (18.8% and 12.6%) found in two types of flax stem waxes. Overall, the predominant long-chain FAs found in the outer layer and fibre bundles were palmitic and stearic acids, with 70-80% present in the outer layer (Morrison and Akin 2001). In another study conducted by Guttierrez and Del Rio (2003), the most predominant lipids in flax fibres were waxes, accounting for 74%, and were constituted of long-chain n-fatty acids esterified with long-chain n-fatty alcohols. The principal FAs associated with waxes were mixtures of C16–C30 FAs, with C18 FA being the single most abundant FA component. The minor difference in wax and cutin material composition reported in these independent studies may be accounted for by the efficacy of removal of the outer layer from the fibre during processing (Morrison and Akin 2001). Regardless, cutin and wax lipids deserve attention because they play important roles in plant biology and also because they have important industrial applications such as coating and in the cosmetics industry.

1.2.2. Leaves

Leaves are not only the factory for photosynthesis, capturing the light flux and synthesizing primary carbon metabolites, but also contain significant amounts of primary and secondary metabolites playing crucial roles in plant signalling and defence (Fofana et al. 2002, 2005, McNally et al. 2003a, 2003b). The leaves of higher plants are a rich source of FAs, particularly linoleate and linolenate (Wharfe and Harwood 1978). Murphy and Leech (1981) provided the first evidence that the chloroplast was able to synthesize all of the saturated FAs required by the leaf and that collaboration of other organelles was recruited for the production of FA precursors (acetate) and FA desaturation steps. Indeed, during glycolysis, carbohydrates provide substrates for acetyl-CoA which is rate limiting in FA biosynthesis. Over a decade ago, Ohlrogge and Browse (1995) showed that the synthesis of leaf membrane lipids occurs through a cooperation between the chloroplasts and the endoplasmic reticulum. Flax total lipids extracted from mature leaves (35 day old) consist of neutral lipids, galactolipids and phospholipids (Chechetkin et al. 2008, 2009). Recently, an enzyme preparation from flax leaves revealed lipoxygenase, divinyl ether synthase and allen oxide synthase activity against linoleic and α-linolenic acids and their respective 13-hydroperoxides in vitro, and an unprecedented family of complex oxylipins were detected in the 35 day old flax leaves (Chechetkin et al. 2008). Oxylipins are a large family of bioregulators synthesised via unsaturated FA. In flax leaves, two major oxylipins, linolipins A and B, absorbing λ_{max} at 267 nm, were identified and characterized (Chechetkin et al. 2008). Transesterification products of linolipin A consisted of the methyl esters of α-linolenic acid and (ω5Z)-etherolenic acid whereas that of linolipin B consisted of a single transesterification product, namely (ω5Z)-etherolenic acid methyl ester. These linolipins are age-dependent, absent in young leaves (14-23 day old) and seeds and were found to be pathogen-inducible (Chechetkin et al. 2008).

1.2.3. Flower

Flax flower is a composite structure made of sepals, petals, ovary, pistil, filaments and stamens. Gibble and Kurtz (1956) observed for the first time the synthesis of long-chain FAs through multiple condensation of acetate by floating flax flowers on an acetate-1-14C solution. Palmitic, stearic, oleic, linoleic and linolenic acids were formed in developing fruits following multiple condensation of acetate. Recently, Fofana et al. (2004) profiled the FA composition in flax ovary at anthesis. Palmitic, stearic, oleic, linoleic and linolenic acids represented 22.5, 3.4, 3.3, 35.8 and 35.0 % of the total FA composition, respectively. The study also suggested a high turnover of palmitic acid at this early stage and a high desaturation rate by desaturation enzymes stearoyl-ACP desaturase (SAD) and fatty acid desaturase (FAD). It is now well established that the regulation of anthesis and flower development is under the control of plant hormones such as jasmonic acid and oxylipins for which linolenic acid acts as a precursor (Piffanelli et al. 1997, Ishiguro et al. 2001, Suzuki et al. 2003). Moreover, almost all flower parts are covered by cuticular waxes and contain other lipid-based compounds that protect against dessication and pathogens, facilitate pollen-stigma interactions, pollen tube growth and fertilization (Dowd et al. 2006, Helling et al. 2006).

1.2.4. Seeds

1.2.4.1. Features of Seed Fatty Acids

Seed storage lipids are the major source of edible vegetable oils, the quality of which is dependent primarily on their FA composition (Green 1986b). Similarly to many other oilseeds, lipids accumulate in flax seeds as lipid bodies of TAGs, which are formed by an extension of the membrane-lipid biosynthetic pathway common to all plant tissues. Lipid bodies are pools of TAGs surrounded by a single monolayer membrane (Voelker and Kinney 2001). Contrary to the conserved FA composition of membrane lipids, there is a very high divergence among plant species with regard to seed oil acyl chains and saturation. For example, seed oils from many species belonging to the families of *Araceae*, *Lauraceae*, *Lythraceae* and *Ulmaceae* often contain saturated acyl chains ranging from C8 to C14 (Voelker and Kinney 2001) whereas *Brassica* species preferentially deposit elongated chains of C16 – C24. FAs stored in plant seeds are usually unbranched compounds with an even number of carbons ranging from 12 to 22 and 0 to 3 double bonds (Thelen and Ohlrogge 2002) although C14 to C18 homologues are encountered in appreciable concentrations in glycerolipids of most plant species. Thus, plants synthesize more than 200 types of FAs having a specialized nomenclature (Table 2). Many variations are encountered in nature with regard to functional groups such as hydroxy, epoxy, cyclopropene or acetylenic (Thelen and Ohlrogge 2002) rendering the naming of FAs more complex.

Table 2. Systematic, common and shorthand names for plant fatty acids[a]

Systematic name	Common name	Shorthand
Saturated FAs		
Ethanoic	acetic	2:0
Butanoic	butyric	4:0
Hexanoic	caproic	6:0
Octanoic	caprylic	8:0
Decanoic	capric	10:0
Dodecanoic	lauric	12:0
Tetradecanoic	myristic	14:0
Hexadecanoic	palmitic	16:0
Octadecanoic	stearic	18:0
Eicosanoic	arachidic	20:0
Docosanoic	behenic	22:0
Monoenoic FAs		
cis-9-hexadecenoic	palmitoleic	16:1(n-7)
cis-6-octadecenoic	petroselinic	18:1(n-12)
cis-9-octadecenoic	oleic	18:1(n-9)
cis-11-octadecenoic	*cis*-vaccenic	18:1(n-7)
cis-13-docosenoic	erucic	22:1(n-9)
cis-15-tetracosenoic	nervonic	24:1(n-9)
Polyunsaturated FAs		
9,12-octadecadienoic	linoleic	18:2(n-6)
6,9,12-octadecatrienoic	γ-linolenic	18:3(n-6)
9,12,15-octadecatrienoic	α-linolenic	18:3(n-3)
5,8,11,14-eicosatetraenoic	arachidonic	20:4(n-6)

[a] Adapted from Cahoon et al. (2009) and http://www.lipidlibrary.co.uk/Lipids/whatlip/index.htm#fatty

1.2.4.2. Fatty Acid Composition of Flax Seeds

Flax seed FA composition is now generally well known (Green 1986a, 1986b, Siemens and Daun 2005, Yurenkova et al. 2005, Fofana et al. 2006, El-Beltagi et al. 2007, Sebei et al. 2007). The oil content of oilseed and fibre flax was reported to range from 31.4 to 45.7% of dry seed weight (Diederichsen and Raney 2006) but current linseed varieties can produce as much as 50% oil in favourable environments (see section 2.6.1), and display a wide range of FA composition (Table 3). For example, conventional linseed varieties have 50-59% ALA, solin's ALA content is 2-3% while some flax lines (called "high-lin") have as much as 70-73% ALA. Besides the major FAs reported in table 3 for cultivated flax, seeds of related *Linum* species also display high levels of variations with regards to FA profile and content (Yurenkova et al. 2005). Minor FAs such as capric (C10:0), lauric acid (12:0), myristic (C14:0), myristoic (14:1), pentadecanoic (C15:0), palmitoleic (16:1), arachidic (C20:0), eicosenoic (C20:1), eicosadienoic (C20:2), eicosatrienoic (C20:3) and behenic (C22:0) acids have been found in trace or low amounts in most cultivated and related *Linum* species (Yurenkova et al. 2005).

Table 3. Seed fatty acid composition in several cultivated oilseed and fibre flax

Variety	16:0	18:0	18:1	18:2	18:3	22:0	References
AC McDuff	5.9	3.8	18.4	17.9	54	-	Fofana et al. 2006
SP2047	6.3	3.4	16.7	71.5	2.0	-	Fofana et al. unpublished
Linola™	6.2	3.4	16	72	2.6	-	Fofana et al. unpublished
AC Emerson	10.0	8.0	30.0	12.0	40.0	-	Sorensen et al. 2005
Vimy	6.0	4.0	30.0	12.0	48.0	-	Sorensen et al. 2005
Fibre flax	5.7	4.7	28.0	13.5	46.3	0.28	Yurenkova et al. 2005

Of particular interest is the relative proportion of the C18 unsaturated FAs namely oleic acid (C18:1), linoleic acid (C18:2) and linolenic acid (C18:3). These FAs determine the end use of the oils: those with high linoleic acid content like canola, soybean, sunflower, safflower or solin-type flax are utilized in polyunsaturated oils and margarines while those with high oleic acid content, such as peanut and olive oils, are more suitable for cooking and as salad dressing oils (Brockerhoff and Yurkowski 1966). In contrast, oils having high levels of linolenic acid are unsuitable for food use due to problems of flavour reversion associated with auto-oxidation and rancidity (Green 1986b). Release or extraction of the highly unsaturated oil from the seeds and contact with air initiate the rancidity process. For this reason, high ALA flax seeds should preferentially be ground or crushed just prior to usage and flax oil should be kept sealed and refrigerated.

1.2.4.3. TAG Composition of Flax Seeds

As mentioned earlier, seed storage lipids are the main source for edible vegetable oils and are accumulated as TAGs in lipid bodies. Brockerhoff and Yurkowski (1966) performed stereospecific analyses of several vegetable fats. In flax oil, the averaged incorporation rate of unsaturated FAs at the three (the two α and one β) positions of TAG was higher (16, 17, and 57% for 18:1, 18:2 and 18:3, respectively) than that of saturated FAs (6 and 4% for 16:0 and 18:0, respectively). These data are an indication that the more a FA is synthesized, the more likely it will be found in the oil as a TAG by the sequential acylation of glycerol-3-phosphate by glycerol-3-phosphate acyltransferase (GPAT), acyl CoA:LPA acyltransferase (LPAAT) and diacylglycerol acyltransferase (DGAT) (Sorensen et al. 2005). Development of new analytical technologies has increased the efficiency and ease of detecting more TAGs and their FA composition. For example, Krist et al. (2006) have identified more than 20 TAGs directly from whole linseed oil samples using Maldi-RTOF-MS and ESI-IT-MS.

1.3. GENETIC CONTROL, BIOSYNTHESIS AND REGULATION OF FLAX SEED LIPIDS

1.3.1. Genetic Control of Polyunsaturated Fatty Acids (PUFAs)

The genus *Linum* comprises approximately 200 species (Diederichsen 2007). Flax (*Linum usitatissimum* L.) is a diploid (2n = 30) annual self pollinated species. Its genetic diversity and oil content variability have been extensively investigated (Green and Marshall 1981, Diederichsen 2001, Allaby et al. 2005, Fu 2005, Diederichsen and Raney 2006). A positive association between seed oil concentration, yellow seed coat color and seed weight has been reported, making it possible to increase seed oil concentration by combining yellow seed coat color with high seed weight (Diederichsen and Raney 2006). The genetic control of palmitic, palmitoleic and PUFAs in flax seed oil has been the focus of extensive investigations particularly in Canada and Australia through mutant line screening (Green 1986b, Rowland and Bhatty 1990, Ntiamoah et al. 1995, Ntiamoah and

Rowland 1997, Saeidi and Rowland 1997, Rowland et al. 2003). Palmitic and palmitoleic acid characteristics in flax seeds were found to be the result of the pleiotropic effect of a single additive gene whereas deficiency in linolenic acid was controlled by two independent recessive genes (Green 1986b, Ntiamoah et al. 1995). Despite flax oil's economic importance, it is only recently that major progress has been made in the identification of key genes controlling FA biosynthesis in flax. Singh et al. (1994) and Jain et al. (1999) reported the cDNA and two promoter sequences for *sad*. The two *sad* genes, *sad1* and *sad2*, were paralogous loci harbouring differentially expressed genes. Later, Fofana et al. (2004) cloned the FA biosynthetic genes beta-keto acyl CoA synthase (*kas*), fatty acid elongase (*fae*), *sad* and *fad2*. Two closely related *fad2* sequences were identified based on Southern blotting and sequence homology (Fofana et al. 2004, Khadake et al. 2009). An important accomplishment in this field was made by Vrinten et al. (2005) who identified the two genes *Lufad3A* and *Lufad3B* that encode microsomal desaturases capable of desaturating linoleic acid to form linolenic acid. More details on these gene discoveries can be found in section 2.1. The final acylation of FAs to glycerol backbone to form TAG is controlled by the *dgat1* gene. Although the enzyme has been cloned in several oilseeds including canola (Nykiforuk et al. 2002), soybean (Wang et al. 2006) and castor bean (He et al. 2004), no flax *dgat1* gene has been reported to date. With flax sequencing projects in progress in several laboratories in Canada, France and India, it is expected that more important FA biosynthetic genes and their regulatory elements will be discovered in the near future (see section 2.5).

1.3.2. Biosynthesis and Regulation of PUFAs

The lipid biosynthesis pathway can be conceptually divided into two blocks: one for FA biosynthesis and one for lipid assembly (Weselake et al. 2008). FA biosynthesis in oilseeds is catalyzed by a set of enzymes located in plastids and the series of reactions necessary for de novo synthesis of FAs, up to 18 carbons in length, have been described and elucidated in plants (Voelker and Kinney 2001, Thelen and Ohlrogge 2002). The synthesis of short to long saturated acyl chains (C4-C18) involves the condensation of C2 units from malonyl-acyl carrier proteins to acyl chains and is achieved sequentially by KAS enzymes I, II, III and IV according to their substrate specificity (Voelker and Kinney 2001). KASIII initiates the FA synthesis in plants by catalyzing the condensing reaction of acetyl-CoA and malonyl-

ACP to form 3-ketobutyryl-ACP (Tai and Jaworski 1993). Subsequent condensation reactions are catalyzed by other members of the KAS family, namely KASI, II and IV (Shimakata and Stumpf 1982, Kauppinen et al. 1988, Dehesh et al. 1998). Unlike KASIII, these other condensing enzymes utilize only acyl-ACPs for elongation by malonyl-ACP. Biochemical analyses of the plant condensing enzymes indicate that KASI catalyzes the majority of the condensation steps using acyl-ACPs as substrates, up to 14:0, whereas KASII functions primarily in stearic acid synthesis preferring C14:0 and C16:0 as substrates (Shimakata and Stumpf 1982, Dehesh et al. 2001).

During lipid biosynthesis, desaturations of FAs are important steps in oil biochemistry determining the ratio of total saturated to unsaturated FA and the end-use properties of the oils (Shanklin and Somerville 1991, Knutzon et al. 1992, Mikkilineni and Rocheford 2003). The first FAs desaturation step is catalyzed by a plastidial SAD (Thelen and Ohlrogge 2002, Iba et al. 1993). Recent studies showed that *kas*, *sad* and *fad2* are expressed early during seed development. *Sad* and *fad2* appeared to be transcriptionally and co-ordinately regulated and *fad2* seemed to be the most sensitive to environmental conditions in the field (Fofana et al. 2004, 2006). Manipulation of these genes to produce designer flax with optimized FA composition should take into consideration the roles played by these enzymes in other tissues, such as ovaries, as they may affect other essential biosynthetic pathways.

After termination of the plastidial FA elongation by acyl-ACP thioesterases which hydrolyze acyl chains from ACP, the free FAs are channelled into the cytoplasm, where they are subjected to further elongation and desaturation steps on the endoplasmic reticulum. These steps are catalyzed respectively by a FAE and several FADs leading to ω-6 and ω-3 PUFAs (Miquel and Browse 1994, Schultz et al. 2000, Han et al. 2001, Voelker and Kinney 2001). The elongation of acyl pools by FAE is limited. Although highly expressed in transgenic flax lines, Abbadi et al. (2004) observed a strong bottleneck in efficiently producing very long chain PUFAs (VLCPUFAs) in TAGs. In developing seeds of oilseed plants, DGAT catalyses the acyl-coenzyme A dependent acylation of *sn*-1,2-diacylglycerol (DAG) to generate TAG (Weselake et al. 2008). It may therefore be possible to produce eicosapentaenoic acid (EPA) directly in flax seed if adequate elongase and desaturase activity using C18:3 as substrate can be achieved to eliminate or alleviate the bottleneck encountered by the endogenous FAEs and described by Abbadi et al. (2004) (see section 2.6).

Chapter 2

CURRENT TRENDS AND FUTURE PERSPECTIVES IN APPLIED GENOMICS AND TRANSGENIC RESEARCH IN FLAX

2.1. TRANSCRIPTOMICS AND GENE DISCOVERY

One of the resources of GenBank namely, dbEST (Boguski et al. 1993) contains sequence data and annotation details of cDNA sequences (Expressed Sequence Tags or ESTs) from many organisms including flax. This dbEST resource is used to identify important genes expressed in specific developmental stages/tissues. For flax, ~7969 EST sequences were available (NCBI, 2 November 2009), generated primarily from fibre bearing stem tissues and developing bolls. Similarly, the gene discovery program of the NAtural Product GENomic consortium (NAPGEN) of Canada generated ~140,591 ESTs originating from the embryo, endosperm and seed coat of flax cultivar CDC Bethune (Cloutier et al. 2009). Using the knowledge of the key genes of the FA biosynthetic pathway in heterologous species namely, *Helianthus* sp, *Brassica* sp and soybean, the genes encoding the condensing enzymes KAS and FAE and desaturating enzymes SAD1, SAD2 and FAD2 have been cloned from cDNAs of 2-12 days after anthesis (DAA) developing bolls (Fofana et al. 2004). Vrinten et al. (2005) cloned two *fad3* genes (*Lufad3A* and *Lufad3B*) governing the linolenic acid trait in flax from two cDNA libraries constructed from 15-20 DAA developing embryos of cultivar 'CDC Normandy' and a solin (low linolenic acid) flax line '593-708'. In the solin type, both genes carry point mutations associated with premature stop codons resulting in low linolenic acid (2-3%, Table 3). In cultivar CDC Normandy, the full length cDNA clones corresponding to wild type *fad3A* and *fad3B* were 1475 bp and 1328 bp in lengths with ORFs coding for 392 and 391-amino acids, respectively. The solin *fad3A* consisted of only 291 amino acids while the *fad3B* sequence was predicted to encode a short polypeptide of only 53 amino acids, both caused by single nucleotide polymorphisms creating premature stop codons. In another pioneering study, analysis of expression profiles of *sad* and *fad2* genes in ovary at anthesis and in developing bolls from 2-32 DAA, suggested the presence of transcriptional modulation during the seed development period (Fofana et al. 2006). High expression of *sad* was found in ovaries at anthesis, followed by a decrease in expression at 4 DAA and mild upregulation at 8 DAA. At maturity, low levels of expression at a rate of 2.1% to 4.5% were observed relative to the base level found in ovaries. On the other hand, low expression of *fad2* was found in the early stages of seed development which gradually increased from 8 DAA and peaked at 16 DAA, at which stage, it was estimated to be at 158% of the level of transcripts observed in ovaries. Although *fad2* expression increased from 8 to 16 DAA, a corresponding change in the linoleic acid content was not observed. This indicated that the linoleic acid produced by FAD2 from oleic acid was in a transient state and was acted upon by FAD3 to produce linolenic acid, whose proportion increased steadily from 8 to 16 DAA (Fofana et al. 2006). Also, environmental cues, such as warm and dry weather, negatively influenced the levels of linoleic and linolenic acids, implying reduced expression of *fad2* under such conditions.

A full length *fad2* gene (1137 bp) encoding a 378-amino acid desaturase was also cloned from genomic DNA of flax cultivar Nike (Krasowska et al. 2007). As well, a 1149 bp long *fad2* gene encoding 382 amino acids was isolated from the flax variety NL97 (Khadake et al. 2009). Both genes were successfully expressed in the yeast (*Saccharomyces cerevisiae*) heterologous system and were localized to the endoplasmic reticulum.

2.2. MOLECULAR MARKER SYSTEMS FOR GERMPLASM CHARACTERIZATION

Flax germplasm has been characterized by marker systems such as random amplified polymorphic DNA (RAPD; Stegnii et al. 2000, Fu et al. 2003a, 2003b), restriction fragment length polymorphism (RFLP; Oh et al. 2000), amplified fragment length polymorphism (AFLP; Spielmeyer et al. 1998, Everaert et al. 2001, Adugna et al. 2006) and simple sequence repeats

(SSRs; Wiesnerová and Wiesner 2004, Roose-Amsaleg et al. 2006, Cloutier et al. 2009). Such markers were used to assess genetic diversity, mainly for discrimination of accessions to be used in flax breeding programs and also to provide unique descriptors to meet distinctness, uniformity and stability (DUS) criteria under the union for the protection of plant varieties (UPOV 1995). However, molecular diversity assessed with these limited numbers of markers was not focussed on seed characteristics such as seed oil content and/or FA composition traits (Diederichsen and Raney 2008). Additional limitations to some of these marker systems include: unreliability (RAPD), labour intensive protocols (RFLP, AFLP) and unsuitability for cross-species application (RAPD, AFLP). In contrast, SSR markers, which can be derived from either genomic sequences or expressed sequence tags (EST-SSRs), are an ideal marker system because of their relative abundance, genome-wide distribution and co-dominant nature (Varshney et al. 2005).

In flax, EST-SSRs are being developed for tagging expressed genes and to generate a linkage map (Cloutier et al. 2009). This resource, while constantly expanding, remains limited with only 39 SSR markers reported in two studies focussed on variety identification (Wiesner et al. 2001, Roose-Amsaleg et al. 2006) in addition to the 275 polymorphic EST-SSR loci recently reported (Cloutier et al. 2009).

2.3. LINKAGE MAPS

In flax, only two linkage maps have been published (Spielmeyer et al. 1998, Oh et al. 2000). The first map consists of 15 linkage groups based on 94 restriction fragment length polymorphism (RFLP) and random amplified polymorphic DNA (RAPD) markers, covering ~1,000 cM of the genome (Oh et al. 2000). The linkage groups comprised from four to ten markers. The second map comprised 18 linkage groups with 213 amplified fragment length polymorphism (AFLP) markers covering 1,400 cM of the genome, with a map density of one marker per 10 cM (Spielmeyer et al. 1998). Recently, an integrated map of flax comprising 502 AFLP markers has been developed (Vromans 2006). Cross applicability of these maps is limited because of the nature of the markers and the marker density. The recent development of EST-SSR markers will alleviate some of these difficulties and permit the construction of reliable, cost effective and cross-applicable genetic maps (Cloutier et al. 2009).

2.4. QUANTITATIVE TRAIT LOCUS (QTL) ANALYSIS

Quantitative traits such as seed weight are governed by many genes with relatively small additive effects and high interaction with the environment, resulting in a display of continuous variation which makes them complex characteristics to deal with in classical breeding programs (Asins 2002). The advent of tools such as molecular marker based linkage maps combined with specialized statistical methods assisted in identification of genomic regions associated with quantitative traits (Doerge 2002). However, without a high density molecular map, such QTLs cannot be accurately identified. To date in flax, QTLs have been identified for a single trait, namely, resistance to *Fusarium oxysporum* fsp *lini*, by the use of an AFLP linkage map generated from two parents differing in their susceptibility to the pathogen (Spielmeyer et al. 1998). On the other hand, in oilseed crops like soybean and *Brassica* sp, many QTLs associated with seed oil content have been identified (for review, see Weselake et al. 2009). In these QTL studies, some 13 and 14 genomic regions associated with seed oil content were identified in soybean and *Brassica napus*, respectively (Weselake et al. 2009). The development of high density molecular maps and large segregating populations, combined with accurate phenotypic characterization over multiple environments, will permit similar assessments of QTLs in flax to be performed in the next few years.

2.5. BAC LIBRARIES AND THE FLAX GENOME INITIATIVE

Flax has an estimated genome size of 0.7 pg/1C nucleus (~676 MB) and 35% of the genome is deemed to be highly repetitive (Cullis 2005, 2007). Development of resources, such as large insert BAC libraries (Shizuya et al. 1992) with sufficient genome coverage enable the construction of physical maps often necessary to anchor whole genome sequences of large genomes. Combined with genetic map(s) and phenotypic assessments, BAC libraries are an essential tool for map based cloning of target genes.

Sequencing of the genome of flax variety CDC Bethune using a whole genome shotgun approach, performed with next generation sequencing technologies, is currently being conducted by a Canadian team (www.genomecanada.ca/en/medias/news.aspx?i=330). In this shotgun sequencing strategy, the entire genome is fragmented into pieces of defined sizes and sequence reads are generated from a very large number of fragments to produce highly redundant sequence coverage across the genome. Powerful computational tools are used to assemble the sequence reads (including BAC end sequences) and anchor them to a physical map for the assembly of contigs and scaffolds. Whole genome sequencing of flax will greatly complement the efforts of marker and gene discovery, which promise to accelerate flax improvement through applied genomics. However, a key challenge of applying this strategy for de novo sequencing remains the presence of large repetitive regions.

Physical maps are also essential for sequencing target regions and to understand the structural organization of the genome. In flax, BAC libraries were constructed in our laboratory for two genotypes, CDC Bethune and M5791. The salient features of the libraries are summarised below (Table 4). These two genotypes differ substantially for many attributes such as oil content and FA composition among others. Comparative structural genomics studies are underway using the two libraries.

Table 4. Characteristic features of two flax BAC libraries

Genotype	M5791		CDC Bethune	
BAC vector	pIndigoBAC-5		pIndigoBAC-5	
E. coli host	DH10B		DH10B	
Enzymes	*Hind*III	*Bam*HI	*Hind*III	*Bam*HI
Number of clones	41,472	54,144	40,704	51,456
Average insert size (Kb)	160	150	150	135
Genome coverage	9.5X	11.6X	8.7X	9.9X

2.6. TRANSGENIC PLANTS

Classical breeding methods resulted in the development of rapeseed with low erucic acid (canola, Stefansson et al. 1961), sunflower with high oleic acid (Pérez-Vich et al. 2000) and soybean with high stearic acid (Graef et al. 1985). Considering the empirical nature of conventional selection methods and the time involved in their development; precise genetic manipulation by a transgenic approach could be beneficial. Careful considerations do, however, have to be given to the commercial production of genetically modified flax because such products are currently not widely accepted in a number of markets. Despite a proliferation of gene validation methods such as virus induced gene silencing (VIGS) or heterologous expression, complementation by transformation remains a powerful method for the demonstration of gene function. Alternatives to a transgenic approach for breeding programs may include implementation of marker-assisted selection, QTL, wild allele introgression, gene pyramiding, etc.

The prospects of genetic engineering of oil i.e., more oil yield per se and altered FA composition, should not be dismissed, even though altering the oil metabolism through manipulation of enzymes involved in the biosynthetic process has met with limited success due to the complex nature of the metabolic flux underlying carbon partitioning (Thelen and Ohlrogge 2002, Dyer and Mullen 2005, Napier et al. 2006). A gene encoding an enzyme can be silenced or overexpressed or a new gene encoding an enzyme extending the pathway to yield a novel product can be introduced (Singh et al. 2005, Truksa et al. 2006). Gene silencing can be achieved through an antisense mRNA approach or the more precise RNAi method to knock out the expression of the endogenous gene(s) (Knauf 1987, Liu et al. 2002, Flores et al. 2008, Auer and Frederick 2009). *Agrobacterium* mediated transformation and regeneration of transgenic plants have been established in flax (Jordan and McHughen 1988) and seed specific promoters have been isolated and characterized (Jain et al. 1999, Drexler et al. 2003a) which paved the way to modify the FA profile by metabolic engineering (Abbadi et al. 2004, Truksa et al. 2006).

2.6.1. Increased Oil Content

As mentioned earlier, flax seeds have an oil content averaging 38.3% (Diederichsen and Raney 2006) but current linseed varieties range from 45-50% oil. Increased seed oil content is an important objective in the breeding of oilseed crops, including flax, because oil is the primary product for both human consumption and industrial applications. Advancements in the identification of QTLs with positive additive effects for oil content and marker assisted introgression/pyramiding of such QTLs as well as manipulation of the oil biosynthetic pathway by genetic engineering show promise in crops like soybean, *Brassica* sp and maize (Weselake et al. 2009). Briefly, oil biosynthesis occurs via the Kennedy pathway which starts when FAs from the acyl-CoA pool are incorporated into the glycerol backbone (Ohlrogge and Browse 1995). In the first step, GPAT catalyzes the acylation of glycerol-3-phosphate (G3P), to yield lysophosphatic acid (LPA), which is converted into phosphatidic acid (PA) by LPAAT. Using PA as substrate, *sn*-1,2-diacylglycerol (DAG) is produced by the action of phosphatase. Final acylation of DAG is carried out by an acylCoA:diacylglycerol acyltransferase (DGAT) to yield TAG (Cases et al. 1998, Zhang et al. 2005). Two minor acyl-CoA independent pathways of oil biosynthesis are also available in plants (Weselake et al. 2009). Biochemical analysis identified the final acylation step mediated by the DGAT as the rate-limiting step in oil synthesis and accumulation in plants making this enzyme a target for increased oil content (Lung and Weselake 2006).

Brassica napus cv Hero transformed with the yeast *sn*-2 acyltransferase (*slc1-1*) gene consistently yielded more oil and a higher proportion of erucic acid (C22:1) than traditional industrial rapeseed cultivars (Zou et al. 1997, Taylor et al. 2001). Weselake et al. (2008) have shown an increase of 14% in seed oil content in canola lines overexpressing the *dgat1* gene. Similarly, positional cloning of a QTL governing high maize seed oil content and high proportion of oleic-acid identified a variant of the acyl-CoA:diacylglycerol acyltransferase referred to as '*dgat1-2*' (Zheng et al. 2008). Ectopic expression of the *dgat1-2* allele increased oil content by 41% and oleic acid proportion by 107% (Zheng et al. 2008), providing further evidence that targeting DGAT activity has potential for increasing oil content. Recently, Taylor et al. (2009) have cloned *dgat1* genes from both *A. thaliana* and *B. napus* cv Jet Neuf and overexpressed them in canola seed. The *A. thaliana dgat1* gene was inserted into *B. napus* cv Quantum whereas a doubled haploid line (DH12075) from *B. napus* was transformed with the gene from *B. napus* cv Jet Neuf. Both transgenics have shown an increase in seed oil content by 2.5-7% under field conditions (Taylor et al. 2009). Hence, the prospect of isolating *dgat* orthologs from the flax genome and germplasm and understanding the metabolic flux (Bates et al. 2009) could pave the way to improved oil content.

2.6.2. Modification of Fatty Acid Composition for Food Uses

There is increased demand for nutritionally important ω-3 FAs namely, ALA (C18:3[9, 12, 15]), EPA (C20:5[5, 8, 11, 14, 17]) and docosahexaenoic acid (DHA, C22:6[4,7,10,13,16,19]), due to dwindling fish resources and the high cost of extraction from marine microalgae (Myers and Worm 2003, Truksa et al. 2009). Humans cannot synthesize C18 FAs which, on their own and upon elongation to EPA and DHA, play a crucial role as membrane phospholipids especially in the brain, the fetal neuronal system and in cardiovascular health. Hence ω-3 FAs must be supplemented through diet (Damude and Kinney 2008). Dietary intake of flax seed/oil represents an excellent source of ALA. Linoleic and linolenic acids function as substrates for further elongation to produce VLCPUFAs such as arachidonic acid (ARA, C20:4 [5, 8, 11, 14], ω-6 FA), EPA and DHA. Though humans can synthesize VLCPUFAs from the dietary EFAs (C18 group); infants, pregnant women and the elderly have an increased demand for VLCPUFAs (Sands et al. 2009). To meet the needs of these special groups of people, the following options are available. First, eating a balanced diet with foods rich in EPA and DHA, such as marine fishes and supplementary oils from algae commercially available in markets. Second, consumption of flax which has a good ω-3:ω-6 ratio favourable for conversion into EPA and DHA in humans, especially with a vegetarian diet. A third option is the production of VLCPUFAs in plants themselves by introducing the enzymatic systems involved in the biochemical pathway necessary for conversion of ω-3 and ω-6 available in the host species.

To date, biochemically characterized plants, including major oilseed crops, lack the genes for the synthesis of EPA and DHA and therefore the genes involved in the biosynthesis of VLCPUFAs in marine microalgae and fungi have been used to produce transgenic plants (Napier 2007, Truksa et al. 2009). Transgenic flax with heterologous genes has been reported to be enriched with VLCPUFA, though the level was much lower than the level found in microbes (Abbadi et al. 2004). In this study, accumulations of ARA and EPA up to 1.5% were reported in plants transformed with three genes involved in VLCPUFA biosynthesis (Δ6 desaturase, Δ6-elongase and Δ5 desaturase) with seed specific promoters. However, the level of ~12% observed in fish oil has not been achieved due to constraints in the substrate availability for the elongation pathway. *Brassica juncea* transgenics expressing 5-6 heterologous genes involved in the biosynthesis of VLCPUFAs, produced up to 25% and 15% of ARA and EPA, respectively (Wu et al. 2005). The addition of three more heterologous genes (an elongase with specificity for 18- and 20-C FAs, a Δ4 desaturase, and a lysophosphatidic acid acyltransferase) produced lines with DHA levels increased from 0.2 to 1.5% of total FAs. Elongation efficiency of EPA into DHA was found to be only 4% and this bottleneck was attributed to non release of EPA into the acyl-CoA pool (Wu et al. 2005). Similarly, construction of a VLCPUFA pathway in soybean by stacking seven heterologous genes namely, *Arabidopsis* Δ15 desaturase, *Saprolegnia diclina* C20-specific Δ17 desaturase, *S. diclina* Δ6 desaturase, *Mortierella alpina* Δ5 desaturase, *Schizochytrium aggregatum* Δ4 desaturase and ELO-type condensing enzymes from *M. alpina* and *Pavlova* sp. yielded increased EPA levels up to 20% and ~3% DHA in seed (Kinney et al. 2004, Clemente and Cahoon 2009). Recently, seed specific expression of two novel genes namely, *CpDesX* from *Claviceps purpurea* encoding an 18-carbon ω3 desaturase and *Pir-ω3* from *Pythium irregulare* encoding a 20-carbon ω3 desaturase in the host of zero-erucic acid *Brassica carinata* resulted in increased EPA levels up to 25% (Cheng et al. 2009). Further studies are needed to address the bottleneck of substrate availability for the elongation pathway and alternative strategies for the reconstitution of the pathway leading to biosynthesis of VLCPUFAs including DHA, but flax may turn out advantageous because of its intrinsic high level of ω-3 and substantially different ratio of ω-3:ω-6 as compared to these other species.

2.6.3. Modification of the Fatty Acid Profile for Industrial Purposes

Unusual FAs with additional functional moieties such as hydroxyl groups have tremendous importance as industrial chemicals and are mainly derived from crude oil (Drexler et al 2003b). Ricinoleic acid (C18:1-OH) obtained from castor bean (*Ricinus communis*) is the only major commercial alternative feedstock in the production of ~100 petrochemicals which are used in the manufacture of lubricants, nylon, encapsulants, emulsifiers and plastic films (Burgal et al. 2008). Ricinoleic acid is synthesized from the substrate of oleic acid (C18:1) by a hydroxylase enzyme (fatty acid hydroxylase 12-FAH12, Galliard and Stumpf 1966) similar to FAD2 at the amino acid level (van de Loo et al. 1995). Ricinoleic acid synthesized from oleic acid is stored in esterified form with a membrane bound lipid namely, phosphatidylcholine (Bafor et al. 1991). Though it is known that ricinoleic acid is released from PC before incorporation into TAG, the pathway by which hydroxyl FA rich TAG is assembled is not known (Dauk et al. 2009). Castor bean, with ~90% ricinoleic acid is not only adapted to tropics but is also not widely grown as an agricultural crop (Atsmon 1989). Castor bean also contains toxins (ricin, RCA_{120}, ricinine) and allergenic proteins and therefore either targeted silencing of the genes involved in their biosynthesis or reconstitution of the castor oil biosynthetic pathway into other suitable hosts with genomic resources has been suggested (Dyer and Mullen 2005, Auld et al. 2009). Thus, engineering the ricinoleic biosynthetic pathway into a widely grown oil seed crop has commercial significance (Thelen and Ohlrogge 2001). Though C18 FA-rich hosts such as *Brassica* species, soybean and flax can all be genetically modified (for review see Dyer and Mullen 2005), flax is an ideal system since linseed oil already has tremendous importance as an industrial feedstock and thereby existing processing capabilities can be used. Soybean on the other hand, is used for both protein and oil purposes and the negative correlation between protein and oil content may be detrimental (Clemente and Cahoon 2009).

Arabidopsis mutants deficient in three major enzymes involved in the biosynthesis of TAG namely, phospholipid:diacylglycerol acyltransferase-1, -2 as well as diacylglylcerol acyltransferase, can show accumulation of hydroxy FA rich TAG when transformed with oleate 12-hydroxylase cDNA which indicated that these three major enzymes do not play a role in the incorporation of hydroxy FAs into TAG (Dauk et al. 2009). Transgenic flax lines to produce oil enriched with hydroxyl FAs for use as a chemical feedstock, could be produced in the near future provided the bottlenecks in the metabolic flux among the substrate pools are understood more thoroughly (Cahoon et al. 2007, Truksa et al. 2009).

2.7. TILLING AND EcoTILLING FOR FLAX IMPROVEMENT

Mutation breeding is a well established method in flax improvement resulting in the development of cultivars with reduced linolenic acid (Green and Marshall 1984, Green 1986c, Rowland et al. 1995, Ntiamoah and Rowland 1997). Targeting Induced Local lesions in Genomes (TILLING), a reverse genetic method which combines traditional mutagenesis with knowledge about genes gained from DNA sequences generated by genome projects and high throughput genotyping capabilities for mutation detection, has been emerging as an important tool in functional genomics and crop improvement (McCallum et al. 2000a, Stemple 2004, Slade and Knauf 2005). Since EMS induced mutagenesis generates a wide range of mutant alleles, primarily by point mutations, the availability of DNA sequences with knowledge of target genes and high throughput genotyping methods aid in polymorphism detection (Comai and Henikoff 2006). TILLING has been found to be useful in barley (Caldwell et al. 2004), wheat (Slade et al. 2004, Dong et al. 2009), maize (Till et al. 2004), rice (Till et al. 2007) and soybean (Cooper et al. 2008) for detecting novel alleles of target loci. Recent studies in maize (Zheng et al. 2008) and in *Tropaeolum majus* (Xu et al. 2008) suggested that a single amino acid substitution can significantly alter the

DGAT1 efficiency, resulting in increased seed oil content. Hence, TILLING would be an ideal tool to identify more efficient alleles of genes involved in the FA/TAG biosynthetic pathways in oilseed crops including flax (Weselake et al. 2009). Unlike RNAi and transposon insertional mutagenesis, TILLING is suitable for identification of phenotypic variants without introduction of foreign DNA. Also, mutant populations can be used in forward genetic strategies once phenotypic characterization is completed for important target traits. It has been suggested that for genome-wide TILLING, a collection of 10,000 reference plants would be ideal for obtaining mutations from a single primer pair per gene (McCallum et al. 2000b). In wheat, 246 alleles have been identified for the homoeologs of the waxy gene in 1920 individuals and 94 of these alleles were predicted to impact the gene product (Slade et al. 2004). Since the genes involved in the FA/lipid biosynthetic pathway are cloned and sequenced in flax, high-throughput mutation identification will capture novel alleles for such genes. Natural genetic variation primarily arising from point mutations for any target gene of interest can be detected by screening flax germplasm collections using the EcoTILLING procedure (Comai et al. 2004). As a component of the Genome Canada project aimed at generating genomic resources for flax improvement (www.genomecanada.ca/en/medias/news.aspx?i=330), a mutant population of 10,000 individuals was developed and is being characterized for important traits including oil content and composition.

CONCLUSION

In the last decade, the popularity of flax as a food ingredient has increased tremendously. The main factor contributing to this popularity is its high content of ω-3 fatty acids. Important for brain function, these EFAs have also been shown to reduce bad cholesterol and to mitigate the risk of heart diseases amongst other health benefits. The same characteristic of high level of unsaturated FAs is also the desired trait for industrial applications in stains, paints and linoleum flooring for example. In this review, we described the complexity of the lipid compounds in flax by outlining the various classes, describing their biosynthetic processes and providing state of the art knowledge of the genetic mechanisms known to be involved in lipid production in a number of pathways with an emphasis on seed storage lipids.

Flax, with more than 55% ALA, is unique among crops. In this era of genomics, this unique factory offers much potential. The development of molecular knowledge and understanding of genetic mechanisms offers unique opportunities that may not be achievable in other crops. Improved oil content and production of waxes with unique properties are certainly two examples of interest. Production of VLCPUFAs and development of very high linolenic acid content lines represent tangible applied genomics targets. Flax is a diploid plant with a relatively small genome. Leaps in technology advancement and current knowledge of other oilseed crops promise to accelerate discoveries in flax and the entire field of applied genomics to crop improvement.

REFERENCES

Abbadi A, Domergue F, Bauer J, Napier JA, Welti R, Zahringer U, Cirpus P, Heinz E (2004) Biosynthesis of very-long-chain polyunsaturated fatty acids in transgenic oilseeds: constraints on their accumulation. *Plant Cell.* 16:2734-2748

Adugna W, Labuschagne MT, Viljoen CD (2006) The use of morphological and AFLP markers in diversity analysis of linseed. *Biodiversity Conservation.* 15:3193-3205

Allaby RG, Peterson GW, Merriwether DA, Fu YB (2005) Evidence of the domestication history of flax (*Linum usitatissimum* L.) from genetic diversity of *SAD2* locus. *Theor. Appl. Genet.* 112:58-65

Asins MJ (2002) Present and future of quantitative trait locus analysis in plant breeding. *Plant. Breed.* 121:281-291

Atsmon D (1989) Castor. In: Oil crops of the world, their breeding and utilization. Robbelen G, Downey RK, Ashri A (eds), McGraw-hill, New York. pp 438-447

Auer C, Frederick R (2009) Crop improvement using small RNAs: applications and predictive ecological risk assessments. *Trends Biotech.* 27:644-651

Auld DL, Zanotto MD, McKeon T, Morris JB (2009) Castor. In: Oil Crops, Handbook of Plant Breeding 4, Vollmann J, Rajcan I (eds), Springer, New York, pp 317-332

Baertschi SW, Ingram CD, Harris TM, Brash AR (1988) Absolute configuration of cis-12-oxophytodienoic acid of flaxseed: implications for the mechanism of biosynthesis from the 13(S)-hydroperoxide of linolenic acid. *Biochem.* 27:18-24

Bafor M, Smith MA, Jonsson L, Stobart K, Stymne S (1991) Ricinoleic acid biosynthesis and triacylglycerol assembly in microsomal preparations from developing castor bean (*Ricinus communis*) endosperm. *Biochem. J.* 280:507-514

Bates PD, Durrett TP, Ohlrogge JB, Pollard M (2009) Analysis of acyl fluxes through multiple pathways of triacylglycerol synthesis in developing soybean embryos. *Plant. Physiol.* 150:55-72

Baur P (1998) Mechanistic aspects of foliar penetration of agrochemicals and the effect of adjuvants. *Recent Research Development in Agricultural and Food Chemistry.* 2:809

Boguski MS, Lowe TMJ, Tolstoshev CM (1993) dbEST-database for 'Expressed sequence tags'. *Nat. Genet.* 4:332-333

Brockerhoff H, Yurkowski M (1966) Stereospecific analyses of several vegetable fats. *J. Lipid. Res.* 7:62-64

Burgal J, Shockey J, Lu C, Dyer J, Larson T, Graham I, Browse J (2008) Metabolic engineering of hydroxy fatty acid production in plants: RcDGAT2 drives dramatic increases in ricinoleate levels in seed oil. *Plant. Biotechnol. J.* 6:819-831

Buschhaus C, Herz H, Jetter R (2007) Chemical composition of epicuticular and intracuticular wax layers on adaxial sides of *Rosa canina* leaves. *Anna. Bot.* 100:1557-1564

Cahoon EB, Shockey JM, Dietrich CR, Gidda SK, Mullen RT, Dyer JM (2007) Engineering oilseeds for sustainable production of industrial and nutritional feedstocks: solving bottlenecks in fatty acid flux. *Curr. Opin. Plant. Biol.* 10:236-244

Cahoon EB, Clemente TE, Damude HG, Kinney AJ (2009) Modifying vegetable oils for food and non-food purposes. In: *Oil Crops, Handbook of Plant Breeding* 4, Vollmann J, Rajcan I (eds), Springer, New York, pp 31-56

Caldwell DG, McCallum N, Shaw P, Muehlbauer GJ, Marshall DF, Waugh R (2004) A structured mutant population for forward and reverse genetics in barley (*Hordeum vulgare* L). *Plant J.* 40:143-150

Cases S, Smith SJ, Zheng Y-W, Myers HM, Lear SR, Sande E, Novak S, Collings C, Welch CB, Lusis AJ, Erickson SK, Farese Jr RV (1998) Identification of a gene encoding an acyl CoA:diacylglycerol acytransferase, a key enzyme in triacylglycerol synthesis. *Proc. Natl. Acad. Sci. U.S.A.* 95:13018-13023

Chapman KD, Tripathy S, Venables B, Desouza AD (1998) *N*-Acylethanolamines: formation and molecular composition of a new class of plant lipids. *Plant. Physiol.* 116:1163-1168

Chechetkin IR, Blufard A, Hamberg M, Grechkin AN (2008) A lipoxygenase-divinyl ether synthase pathway in flax (*Linum usitatissimum* L.) leaves. *Phytochemistry.* 69:2008-2015

Chechetkin IR, Mukhitova FK, Blufard ASB, Yarin AY, Antsygina LL, Grechkin AN (2009) Unprecedented pathogen-inducible complex oxylipins from flax - linolipins A and B. *FEBS J.* 276:4463-4472

Chechetkin IR, Mukhtarova LS, Grechkin AN (2001) Mechanistic aspects of biosynthesis of 12-Oxo-10,15(Z)-Phytodienoic acid and related oxylipins: Effect of pH on cyclization of the oxides of allene (18:3 and 18:2). *Doklady Biochem. Biophys.* 377:125-127

Cheng B, Wu G, Vrinten P Falk K, Bauer J, Qiu X (2009) Towards the production of high levels of eicosapentaenoic acid in transgenic plants: the effects of different host species, genes and promoters. *Transgenic Res.* [Epub ahead of print DOI 10.1007/s11248-009-9302-z]

Clemente TE, Cahoon EB (2009) Soybean oil: Genetic approaches for modification of functionality and total content. *Plant. Physiol.* 151:1030-1040

Cloutier S, Niu Z, Datla R, Duguid S (2009) Development and analysis of EST-SSRs for flax (*Linum usitatissimum* L.). *Theor. Appl. Genet.* 119:53-63

Comai L, Henikoff S (2006) TILLING: practical single-nucleotide mutation discovery. *Plant. J.* 45:684-694

Comai L, Young K, Till BJ, Reynols SH, Greene EA, Codomo CA, Enns LC, Johnson JE, Burtner C, Odden AR, Henikoff S (2004) Efficient discovery of DNA polymorphisms in natural populations by Ecotilling. *Plant. J.* 37:778-786

Cooper JL, Till BJ, Laport RG, Darlow MC, Kleffner JM, Jamai A, El-Mellouki T, Liu S, Ritchie R, Nielsen N, Bilyeu KD, Meksem K, Comai L, Henikoff S (2008) TILLING to detect induced mutations in soybean. *BMC Plant. Biol.* 8:9 doi: 10.1186/1471-2229-8-9

Cullis CA (2005) Mechanisms and control of rapid genomic changes in flax. *Ann. Bot.* 95:201-206

Cullis CA (2007) Flax. In: Kole C (ed) *Genome mapping and molecular breeding in plants.* Springer-Verlag, Berlin Heidelberg, pp 275-295

Damude HG, Kinney AJ (2008) Enhancing plant seed oils for human nutrition. *Plant. Physiol.* 147:962-968

Dauk M, Lam P, Smith MA (2009) The role of diacylglycerol acyltransferase-1 and phospholipids:diacylglycerol acyltransferase-1 and -2 in the incorporation of hydroxy fatty acids into triacylglycerol in

Arabidopsis thaliana expressing a castor bean oleate 12-hydroxylase gene. *Botany.* 87:552-560

Dehesh K, Edwards P, Fillatti J, Slabaugh M, Byrne J (1998) KAS IV: a 3-ketoacyl-ACP synthase from *Cuphea sp.* is a medium chain specific condensing enzyme. *Plant. J.* 15:383-390

Dehesh K, Tai H, Edwards P, Byrne J, Jaworski JG (2001) Overexpression of 3-Ketoacyl-Acyl-Carrier Protein Synthase IIIs in plants reduces the rate of lipid synthesis. *Plant. Physiol.* 125:1103-1114

Diederichsen A (2001) Comparison of genetic diversity of flax (*Linum usitatissimum* L.) between Canadian cultivars and a world collection. *Plant. Breed.* 120:360-362

Diederichsen A (2007) Ex situ collections of cultivated flax (*Linum usitatissimum* L.) and other species of the genus *Linum* L. *Genet. Resour. Crop. Evol.* 54:661-678

Diederichsen A, Raney JP (2006) Seed colour, seed weight and seed oil content in *Linum usitatissimum* accessions held by Plant Gene Resources of Canada. *Plant. Breed.* 125:372-377

Diederichsen A, Raney JP (2008) Pure-lining of flax (*Linum usitatissimum* L.) genebank accessions for efficiently exploiting and assessing seed character diversity. *Euphytica.* 164:255-273

Doerge RW (2002) Mapping and analysis of quantitative trait loci in experimental populations. *Nat. Rev. Genet.* 3:43-52

Dong C, Dalton-Morgan J, Vincent K, Sharp P (2009) A modified TILLING method for wheat breeding. *Plant. Genome.* 2:39-47

Dowd PE, Coursol S, Skirpan AL, Kao T-H, Gilroy S (2006) *Petunia* phospholipase C1 is involved in pollen tube growth. *Plant. Cell.* 18:1438-1453

Drexler HHS, Scheffler JA, Heinz E (2003a) Evaluation of putative seed-specific promoters for *Linum usitatissimum*. *Mol. Breed.* 11:149-158

Drexler H, Spiekersmann P, Meyer A, Domergue F, Zank T, Sperling P, Abbadi A, Heinz E (2003b) Metabolic engineering of fatty acids for breeding of new oilseed crops: strategies, problems and first results. *J. Plant. Physiol.* 160:779-802

Dyer JM, Mullen RT (2005) Development and potential of genetically engineered oilseeds. *Seed. Sci. Res.* 15:255-267

El-Beltagi HS, Salama ZA, El-Hariri DM (2007) Evaluation of fatty acids profile and the content of some secondary metabolites in seeds of different flax cultivars (*Linum usitatissimum* L.). *Gen. Appl. Plant. Physiol.* 33:187-202

Everaert I, Riek JD, Loose MD, Waes JV, Bockstaele EV (2001) Most similar variety grouping for distinctness evaluation of flax and linseed (*Linum usitatissimum*) varieties by means of AFLP and morphological data. *Plant. Var. Seeds.* 14:69-87

Fahy E, Subramaniam S, Brown HA, Glass CK, Merrill AHJ, Murphy RC, Raetz CRH, Russell DW, Seyama Y, Shaw W, Shimizu T, Spener F, van Meer G, VanNieuwenhze MS, White SH, Witzum JL, Dennis EA (2005) A comprehensive classification system for lipids. *J. Lipid Res.* 46:839-862

Fahy E, Subramaniam S, Murphy RC, Nishijima M, Raetz CRH, Shimizu T, Spener F, van Meer G, Wakelam MJO, Dennis EA (2009) Update of the LIPID MAPS comprehensive classification system for lipids. *J. Lipid Res.* 50:S9-S14

Flores T, Karpova O, Su X, Zeng P, Bilyeu K, Sleper DA, Nguyen HT, Zhang ZJ (2008) Silencing of the GmFAD3 gene by siRNA leads to low α-linolenic acids (18:3) of fad3-mutant phenotype in soybean [*Glycine max* (Merr)]. *Transgenic. Res.* 17:839-850

Fofana B, Benhamou N, McNally DJ, Labbé C, Séguin A, Bélanger R (2005) Suppression of induced resistance in cucumber through disruption of the flavonoid pathway. *Phytopathol.* 95:114-123

Fofana B, Cloutier S, Duguid S, Ching J, Rampitsch C (2006) Gene expression of stearoyl-ACP desaturase and Δ12 fatty acid desaturase 2 is modulated during seed development of flax (*Linum usitatissimum*). *Lipids.* 41:705-712

Fofana B, Duguid S, Cloutier S (2004) Cloning of fatty acid biosynthetic genes beta-ketoacyl CoA synthase, fatty acid elongase, stearoyl-ACP desaturase, and fatty acid desaturase and analysis of expression in the early developmental stages of flax (*Linum usitatissimum* L.) seeds. *Plant. Sci.* 166:1487-1496

Fofana B, McNally DJ, Labbé C, Boulanger R, Benhamou N, Séguin A, Bélanger R (2002) Milsana-induced resistance in powdery mildew-infected cucumber plants correlates with the induction of chalcone synthase and chalcone isomerase. *Physiol. Mol. Plant. Pathol.* 61:121-132

Fu YB (2005) Geographic patterns of RAPD variation in cultivated flax. *Crop. Sci.* 45:1084-1091

Fu YB, Guerin S, Peterson GW, Carlson JE, Richards KW (2003a) Assessment of bulking strategies for RAPD analyses of flax germplasm. *Genet. Resource Crop. Evol.* 50:743-746

Fu YB, Rowland GG, Duguid SD, Richards KW (2003b) RAPD analysis of 54 North American flax cultivars. *Crop. Sci.* 43:1510-1515

Galliard T, Stumpf PK (1966) Fat metabolism in higher plants: Enzymatic synthesis of ricinoleic acid by a microsomal preparation from developing *Ricinus communis* seeds. *J. Biol. Chem.* 241:5806-5812

Gibble WP, Kurtz EBJ (1956) The synthesis of long-chain fatty acids from acetate in flax, *Linum usitatissimum* L. *Arch. Biochem. Biophys.* 64:1-5

Graef GL, Fehr WR, Hammond EG (1985) Inheritance of three stearic acid mutants of soybean. *Crop. Sci.* 25:1076-1079

Green AG (1986a) Effect of temperature during seed maturation on the oil composition of low-linolenic genotypes of flax. *Crop. Sci.* 26:961-965

Green AG (1986b) Genetic control of polyunsaturated fatty acid biosynthesis in flax (*Linum usitatissimum*) seed oil. *Theor. Appl. Genet.* 72:654-661

Green AG (1986c) A mutant genotype of flax (*Linum usitatissimum* L.) containing very low levels of linolenic acid in its seed oil. *Can. J. Plant. Sci.* 66:499-503

Green AG, Marshall DR (1981) Variation for oil quantity and quality in linseed (*Linum usitatissimum*). *Aust. J. Agric. Res.* 32:599-607

Green AG, Marshall DR (1984) Isolation of induced mutants in linseed (*Linum usitatissimum*) having reduced linolenic acid content. *Euphytica.* 33:321-328

Guttierrez A, Del Rio JC (2003) Lipids from flax fibers and fate in alkaline pulping. *J. Agric Food Chem.* 51:4965-4971

He X, Turner C, Chen GQ, Lin J-T, McKeon TA (2004) Cloning and characterization of a cDNA encoding diacylglycerol acyltransferase from castor bean. *Lipids.* 39:311-318

Han J, LuhsW, Sonntag K, Zahringer U, Borchardt DS, Wolter FP, Heinz E, Frentzen M (2001) Functional characterization of beta-ketoacyl-CoA synthase genes from *Brassica napus* L. *Plant. Mol. Biol.* 46:229-239

Helling D, Possart A, Cottier S, Klahre U, Kost B (2006) Pollen tube tip growth depends on plasma membrane polarization mediated by tobacco PLC3 activity and endocytic membrane recycling. *Plant. Cell.* 18:3519-3534

Iba K, Gibson S, Nishiuchi T, Fuse T, Nishimura M, Arondel V, Hugly S, Somerville C (1993) A gene encoding a chloroplast ω-3 fatty acid desaturase complements alterations in fatty acid desaturation and chloroplast copy number of the *FAD7* mutant of *Arabidopsis thalina*. *J. Biol. Chem.* 268:24099-24105

Ishiguro A, Kawai-Oda A, Ueda J, Nishida I, Okada K (2001) The defective in anther dehiscence 1 gene encodes a novel phospholipase A1 catalyzing the initial step of jasmonic acid biosynthesis, which synchronizes pollen maturation, anther dehiscence, and flower opening in *Arabidopsis*. *Plant. Cell.* 13:2191-2209

Jain RK, Thompson RG, Taylor DC, MacKenzie SL, McHughen A, Rowland GG, Tenaschuk D, Coffey M (1999) Isolation and characterization of two promoters from linseed for genetic engineering. *Crop. Sci.* 39:1696-1701

Jetter R, Kunst L, Samuels AL (2006) Composition of plant cuticular waxes. In: Riederer M, Muller C (eds) *Biology of the plant cuticle.* Blackwell publishing, Oxford, pp 182-215

Jordan MC, McHughen A (1988) Glyphosate tolerant flax plants from *Agrobacterium* mediated gene transfer. Plant Cell Rep 7:281-284

Kauppinen S, Siggaard-Anderson M, von Wettstein-Knowles P (1988) β-ketoacyl-ACP synthase I of *Escherichia coli*: Nucleotide sequence of the *fabB* gene and identification of the cerulenin binding residue. *Carlsberg. Res. Commun.* 53:357-370

Khadake RM, Ranjekar PK, Harsulkar AM (2009) Cloning of a novel omega-6 desaturase from flax (*Linum usitatissimum* L.) and its functional analysis in *Saccharomyces cerevisiae. Mol. Biotechnol.* 42:168-174

Kinney AJ, Cahoon EB, Damude HG, Hitz WD, Kolar Jr CW, Liu Z-B (2004) Production of very long chain polyunsaturated fatty acids in oilseed plants. WIPO Pub No: WO/2004 /071467; PCT/US2004/005758

Knauf VC (1987) The application of genetic engineering to oilseed crops. *Trends Biotech.* 5:40-47

Knutzon DS, Thompson GA, Radke SE, Johnson WB, Knauf VC, Kridl JC (1992) Modification of *Brassica* seed oil by antisense expression of a stearoyl-acyl carrier protein desaturase gene. *Proc. Natl. Acad. Sci. U.S.A.* 89:2624-2628

Krasowska A, Dziadkowiec D, Polinceusz A, Plonka A, Lukaszewicz M (2007) Cloning of flax oleic fatty acid desaturase and its expression in yeast. *J. Am. Oil Chem. Soc.* 84:809-816

Krist S, Stuebiger G, Bail S, Unterweger H (2006) Analysis of volatile compounds and triacylglycerol composition of fatty seed oil gained from flax and false flax. *Eur. J. Lipid Sci. Technol.* 108:48-60

Liu Q, Singh S, Green A (2002) High oleic and high stearic cottonseed oils: nutritionally improved cooking oils developed using gene silencing. *J. Am. Coll. Nutr.* 21:205S-211S

Lung SC, Weselake RJ (2006) Diacylglycerol acyltransferase: a key mediator of plant triacylglycerol synthesis. *Lipids.* 41:1073-1088

McCallum CM, Comai L, Greene EA, Henikoff S (2000a) Targeted screening for induced mutations. *Nat. Biotechnol.* 18:455-457

McCallum CM, Comai L, Greene EA, Henikoff S (2000b) Targeting Induced Local Lesions In Genomes (TILLING) for plant functional genomics. *Plant. Physiol.* 123:439-442

McNally DJ, Wurms KV, Labbé C, Bélanger RR (2003a) Synthesis of C-glycosyl flavonoid phytoalexins as a site-specific response to fungal penetration in cucumber. *Physiol. Mol. Plant. Pathol.* 63:293-303

McNally DJ, Wurms KV, Labbe C, Quideau S, Bélanger RR (2003b) Complex C-glycosyl flavonoid phytoalexins from *Cucumis sativus. J. Nat. Prod.* 66:1280-1283

Metz JG, Pollard MR, Anderson L, Hayes TR, Lassner MW (2000) Purification of a jojoba embryo fatty acyl-Coenzyme A reductase and expression of its cDNA in high erucic acid rapeseed. *Plant. Physiol.* 122:635-644

Mikkilineni V, Rocheford TR (2003) Sequence variation and genomic organization of fatty acid desaturase-2 (Fad2) and fatty acid desaturase-6 (Fad6) cDNAs in maize. *Theor. Appl. Genet.* 106:1326-1332

Miquel MF, Browse JA (1994) High-oleate oilseeds fail to develop at low temperature. *Plant. Physiol.* 106:421-427

Morrison III WH, Akin DE (2001) Chemical composition of components comprising bast tissue in flax. *J. Agric Food Chem.* 49:2333-2338

Murphy DJ, Leech RM (1981) Photosynthesis of lipids from $^{14}CO_2$ in *Spinacha oleracea. Plant Physiol.* 68:762-765

Myers RA, Worm B (2003) Rapid worldwide depletion of predatory fish communities. *Nature.* 423:280-283

Nandy S, Rowland GG (2008) Dual purpose flax (*Linum usitatissimum* L.) improvement using anatomical and molecular approaches. *Inter. Conf. Flax Other Bast Plants.* pp 31-39

Napier JA (2007) The production of unusual fatty acids in transgenic plants. *Annu. Rev. Plant. Biol.* 58:295-319

Napier JA, Haslam R, Caleron MV, Michaelson LV, Beaudoin F, Sayanova O (2006) Progress towards the production of very long-chain polyunsaturated fatty acid in transgenic plants: plant metabolic engineering comes of age. *Physiologia Plantarum.* 126:398-406

Ntiamoah C, Rowland GG (1997) Inheritance and characterization of two low linolenic acid EMS-induced McGregor mutant flax (*Linum usitatissimum* L.). *Can. J. Plant. Sci.* 77:353-358

Ntiamoah C, Rowland GG, Taylor DC (1995) Inheritance of elevated palmitic acid in flax and its relationship to the low linolenic acid. *Crop. Sci.* 35:148-152

Nykiforuk CL, Furukawa-Stoffer TL, Huff PW, Sarna M, Laroche A, Moloney MM, Weselake RJ (2002) Characterization of cDNAs encoding diacylglycerol acyltransferase from cultures of *Brassica napus* and sucrose-mediated induction of enzyme biosynthesis. *Biochim. Biophys. Acta.* 1580:95-109

Oh TJ, Gorman M, Cullis CA (2000) RFLP and RAPD mapping in flax (*Linum usitatissimum*). *Theor. Appl. Genet.* 101:590-593

Ohlrogge J, Browse J (1995) Lipid biosynthesis. *Plant. Cell.* 7:957-970

Pérez-Vich B, Garcés R, Fernández-Martínez JM (2000) Genetic relationships between loci controlling the high stearic and the high oleic acid traits in sunflower. *Crop. Sci.* 40:990-995

Piffanelli P, Ross JHE, Murphy DJ (1997) Intra-and extracellular lipid composition and associated gene expression patterns during pollen development in *Brassica napus*. *Plant. J* 11:549-562

Pollard M, Beisson F, Li Y, Ohlrogge JB (2008) Building lipid barriers: biosynthesis of cutin and suberin. Trends Plant Sci 13:236-246

Roose-Amsaleg C, Cariou Pham E, Vautrin D, Tavernier R, Solignac M (2006) Polymorphic microsatellite loci in *Linum usitatissimum*. *Mol. Ecol. Notes.* 6:796-799

Rowland GG, Bhatty RS (1990) Ethyl methanesulphonate induced fatty acid mutations in flax. *J. Am. Oil Chem. Soc.* 67:213-214

Rowland GG, McHughen A, Gusta LV, Bhatty RS, MacKenzie SL, Taylor DC (1995) The application of chemical mutagenesis and biotechnology to the modification of linseed (*Linum usitatissimum* L.). *Euphytica.* 85:317-321

Rowland GG, Wilen R, Hormis Y, Saeidi G, Ntiamoah C (2003) Modification of flax (*Linum usitatissimum*) by induced mutagenesis and transformation. International Atomic Energy Agency Technical Documents (IAEA-TECDOCs) 1369:143-149

Saeidi G, Rowland GG (1997) The inheritance of variegated seed color and palmitic acid in flax. *J. Hered.* 88:466-468

Sands DC, Morris CE, Dratz EA, Pilgeram AL (2009) Elevating optimal human nutrition to a central goal of plant breeding and production of plant-based foods. *Plant. Sci.* 177:377-389

Schultz DJ, Suh MC, Ohlrogge JB (2000) Stearoyl-acyl carrier protein and unusual acyl-acyl carrier protein desaturase activities are differentially influenced by ferredoxin. *Plant. Physiol.* 124:681-692

Sebei K, Debez A, Herchi W, Boukhchina S, Kallel H (2007) Germination kinetics and seed reserve mobilization in two flax (*Linum usitatissimum* L.) cultivars under moderate salt stress. *J. Plant. Biol.* 50:447-454

Shanklin J, Somerville C (1991) Stearoyl-acyl-carrier-protein desaturase from higher plants is structurally unrelated to the animal and fungal homologs. *Proc. Natl. Acad. Sci. U.S.A.* 88:2510-2514

Sharmin E, Ashraf SM, Ahmad S (2007) Synthesis, characterization, antibacterial and corrosion protective properties of epoxies, epoxypolyols and epoxy-polyurethane coatings from linseed and *Pongamia glabra* seed oils. *Int. J. Biol. Macromol.* 40:407-422

Shimakata T, Stumpf PK (1982) Isolation and function of spinach leaf β-ketoacyl-[acyl-carrier-protein] synthases. *Proc. Natl. Acad. Sci. U.S.A.* 79:5808-5812

Shizuya H, Birren B, Kim U-J, Mancino V, Slepak T, Tachiiri Y, Simon M (1992) Cloning and stable maintenance of 300-kilobase-pair fragments of human DNA in *Escherichia coli* using an F-factor-based vector. *Proc. Natl. Acad. Sci. U.S.A.* 89:8794-8797

Siemens BJ, Daun JK (2005) Determination of the fatty acid composition of canola, flax, and solin by near-infrared spectroscopy. *J. Am. Oil Chem. Soc.* 82:153-157

Singh S, McKinney S, Green A (1994) Sequence of a cDNA from *Linum usitatissimum* encoding the stearoyl-acyl carrier protein desaturase. *Plant. Physiol.* 140:1075

Singh SP, Zhou X-R, Liu Q, Stymne S, Green AG (2005) Metabolic engineering of new fatty acids in plants. *Curr. Opin. Plant. Biol.* 8:197-203

Slabas AR, Simon JW, Brown AP (2001) Biosynthesis and regulation of fatty acids and triglycerides in oil seed rape. Current status and future trends. *Eur. J. Lipid Sci. Technol.* 103:455-466

Slade AJ, Fuerstenberg SI, Loeffler D, Steine MN, Facciotti D (2004) A reverse genetic, nontransgenic approach to wheat crop improvement by TILLING. *Nat. Biotechnol.* 23:75-81

Slade AJ, Knauf VC (2005) TILLING moves beyond functional genomics into crop improvement. *Transgenic Res.* 14:109-115

Sorensen BM, Furukawa-Stoffer TL, Marshall KS, Page EK, Mir Z, Forster RJ, Weselake RJ (2005) Storage lipid accumulation and acyltransferase action in developing flaxseed. *Lipids.* 40:1043-1049

Spielmeyer W, Green AG, Bittisnich D, Mendham N, Lagudah ES (1998) Identification of quantitative trait loci contributing to Fusarium wilt resistance on an AFLP linkage map of flax (*Linum usitatissimum*). *Theor. Applied. Genet.* 97:633-641

Stegnii VN, Chudinova IV, Salina EA (2000) RAPD Analysis of the flax (*Linum usitatissimum* L) varieties and hybrids of various productivity. *Russ. J. Genet.* 36:1149-1156

Stefansson BR, Hougen FW, Downey RK (1961) Note on the isolation of Rape plants with seed oil free from erucic acid. *Can. J. Plant. Sci.* 41:218-219

Stemple DL (2004) TILLING-a high throughput harvest for functional genomics. *Nat. Rev. Genet.* 5:145-150

Suzuki M, Yamaguchi S, Iida T, Hashimoto I, Teranishi H, Mizoguchi M, Yano F, Todoroki Y, Watanabe N, Yokoyama M (2003) Endogenous-ketol linolenic acid levels in short day-induced cotyledons are closely related to flower induction in *Pharbitis nil*. *Plant Cell Physiol.* 44:35-43

Tai H, Jaworski JG (1993) 3-Ketoacyl-acyl carrier protein synthase III from spinach (*Spinacia oleracea*) is not similar to other condensing enzymes of fatty acid synthase. *Plant. Physiol.* 103:1361-1367

Tang GQ, Novitzky WP, Griffin HC, Huber SC, Dewey RE (2005) Oleate desaturase enzymes of soybean: evidence of regulation through differential stability and phosphorylation. *Plant. J.* 44:433-446

Taylor DC, Katavic V, Zou J, MacKenzie SL, Keller WA, An J, Friesen W, Barton DL, Pedersen KK, Giblin EM, Ge Y, Dauk M, Sonntag C, Luciw T, Males D (2001) Field testing of transgenic rapeseed cv. Hero transformed with a yeast *sn*-2 acyltransferase results in increased oil content, erucic acid content and seed yield. *Mol. Breed.* 8:317-322

Taylor DC, Zhang Y, Kumar A, Francis T, Giblin EM, Barton DL, Ferrie JR, Laroche A, Shah S, Zhu W, Snyder CL, Hall L, Rakow G, Harwood JL, Weselake RJ (2009) Molecular modification of triacylglycerol accumulation by over-expression of DGAT1 to produce canola with increased seed oil content under field conditions. *Botany.* 87:533-543

Thelen JJ, Ohlrogge JB (2002) Metabolic engineering of fatty acid biosynthesis in plants. *Metab. Eng.* 4:12-21

Thole JM, Nielsen E (2008) Phosphoinositides in plants: novel functions in membrane trafficking. *Curr. Opin. Plant. Biol.* 11:620-631

Till BJ, Cooper J, Tai TH, Colowit P, Greene EA, Henikoff S, Comai L (2007) Discovery of chemically induced mutations in rice by TILLING. *BMC Plant. Biol.* 7:19 doi:10.1186/1471-2229-7-19

Till BJ, Reynolds SH, Weil C, Springer N, Burtner C, Young K, Bowers E, Codomo CA, Enns LC, Odden AR, Greene EA, Comai L, Henikoff S

(2004) Discovery of induced point mutations in maize genes by TILLING. *BMC Plant. Biol.* 4:12 doi:10.1186/1471-2229-4-12

Truksa M, Vrinten P, Qiu X (2009) Metabolic engineering of plants for polyunsaturated fatty acid production. *Mol. Breed.* 23:1-11

Truksa M, Wu G, Vrinten P, Qiu X (2006) Metabolic engineering of plants to produce very long-chain polyunsaturated fatty acids. *Transgenic Res.* 15:131-137

van de Loo FJ, Broun P, Turner S, Somerville C (1995) An oleate 12-hydroxylase from *Ricinus communis* L. is a fatty acyl desaturase homolog. *Proc. Natl. Acad. Sci. U.S.A.* 92:6743-6747

Van der Selt M, Noordermer MA, Kiss T, van Zadelhoff G, Merghart B, Veldink GA, J.F.G. V (2000) Formation of a new class of oxylipins from N-acyl(ethanol)amines by the lipoxygenase pathway. *Eur. J. Biochem.* 267:2000-2007

Van Maarseveen C, Han H, Jetter R (2009) Development of the cuticular wax during growth of *Kalanchoe daigremontiana* (Hamet et Perr. de la Bathie) leaves. *Plant, Cell Environ.* 32:73-81

Varshney RK, Graner A, Sorrells ME (2005) Genic microsatellite markers in plants: features and applications. *Trends Biotechnol.* 23:48-55

Voelker T, Kinney AJ (2001) Variations in the biosynthesis of seed-storage lipids. *Ann. Rev. Plant. Physiol. Plant. Mol. Biol.* 52:335-361

Vrinten P, Hu Z, Munchinsky M-A, Rowland GG, Qiu X (2005) Two *FAD3* desaturase genes control the level of linolenic acid in flax seed. *Plant. Physiol.* 139:79-87

Vromans J (2006) Molecular genetic studies in flax (*Linum usitatissimum* L.). Wageningen University, Wageningen, p 144

Wanasundara PKJPD, Wanasundara UN, Shahidi F (1999) Changes in flax (*Linum usitatissimum* L.) seed lipids during germination. *J. Am. Oil Chem. Soc.* 76:41-48

Wang HW, Zhang J-S, Gai J-Y, Chen S-Y (2006) Cloning and comparative analysis of the gene encoding diacylglycerol acyltransferase from wild type and cultivated soybean. *Theor. Appl. Genet.* 112:1086-1097

Weselake RJ, Shah S, Tang M, Quant PA, Snyder CL, Furukawa-Stoffer TL, Zhu W, Taylor DC, Zou J, Kumar A, Hall L, Laroche A, Rakow G, Raney P, Moloney MM, Harwood JL (2008) Metabolic control analysis is helpful for informed genetic manipulation of oilseed rape (*Brassica napus*) to increase seed oil content. *J. Exp. Bot.* 59:3543-3549

Weselake RJ, Taylor DC, Rahman MH, Shah S, Laroche A, McVetty PBE, Harwood JL (2009) Increasing the flow of carbon into seed oil. *Biotech. Adv.* 27:866-878

Wharfe J, Harwood JL (1978) Fatty acid biosynthesis in the leaves of barley, wheat and pea. *Biochem. J.* 174:163-169

Wiesner I, Wiesnerová D, Tejklová E (2001) Effect of anchor and core sequence in microsatellite primers on flax fingerprinting patterns. *J. Agric. Sci.* 137:37-44

Wiesnerová D, Wiesner I (2004) ISSR-based clustering of cultivated flax germplasm is statistically correlated to thousand seed mass. *Mol. Biotechnol.* 26:207-214

Wu G, Truksa M, Datla N, Vrinten P, Bauer J, Zank T, Cirpus P, Heinz E, Qiu X (2005) Stepwise engineering to produce high yields of very long-chain polyunsaturated fatty acids in plants. *Nature Biotechnol.* 23:1013-1017

Xu J, Francis T, Mietkiewska E, Giblin EM, Barton DL, Zhang Y, Zhang M, Taylor DC (2008) Cloning and characterization of an acyl-CoA-dependent *diacylglycerol acyltransferase 1* (*DGAT1*) gene from *Tropaeolum majus*, and a study of the functional motifs of the DGAT protein using site-directed mutagenesis to modify enzyme activity and oil content. *Plant. Biotechnol. J.* 6:799-818

Yurenkova SI, Kubrak SV, Tikok VV, Khotyljova LV (2005) Flax species polymorphism for isozyme and metabolic markers. *Russ J. Genet.* 41:256-261

Zhang F-Y, Yang M-F, Xu Y-N (2005) Silencing of *DGAT1* in tobacco cause a reduction in seed oil content. *Plant. Sci.* 169:689-694

Zheng P, Allen WB, Roesler K, Williams ME, Zhang S, Li J, Glassman K, Ranch J, Nubel D, Solawetz W, Bhattramakki D, Llaca V, Deschamps S, Zhong G-Y, Tarczynski MC, Shen B (2008) A phenylalanine in DGAT is a key determinant of oil content and composition in maize. *Nat. Genet.* 40:367-372

Zohary D (1999) Monophyletic vs. polyphyletic origin of the crops on which agriculture was founded in the Near East. Genet Resour Crop Evol 46:133-142

Zou J, Katavic V, Giblin EM, Barton DL, MacKenzie SL, Keller WA, Hu X, Taylor DC (1997) Modification of seed oil content and acyl composition in the Brassicaceae by expression of a yeast *sn*-2 acyltransferase gene. *Plant. Cell.* 9:909-923

INDEX

A

accounting, 8
acid, 5, 6, 7, 9, 12, 14, 15, 17, 18, 21, 22, 23, 24, 25, 26, 29, 31, 32, 33, 35, 36, 37, 38, 39, 40, 41, 42, 43
acylation, 13, 14, 15, 22
agriculture, 43
alcohols, 3, 5, 7
aldehydes, 5, 7
algae, 24
allele, 22, 23
amines, 5, 42
amino acids, 18
annotation, 17
anther, 36
antisense, 22, 37
Arabidopsis thaliana, 34

B

BAC, 20, 21
barriers, 39
benzene, 3
biochemistry, 15
bioregulators, 9
biosynthesis, vii, 1, 9, 14, 15, 22, 24, 25, 26, 31, 33, 36, 37, 39, 41, 42, 43
biosynthetic pathways, 15, 27
biotechnology, 39
biotic, 3
brain, 23, 29
branching, 7
breeding, 1, 19, 20, 21, 22, 26, 31, 33, 34, 39

C

cambium, 7
cancer, vii
carbohydrates, 9
carbon, 3, 8, 22, 25, 42
carbon atoms, 3
castor oil, 25
category a, 6
cDNA, 14, 17, 26, 36, 38, 40
character, 34
chloroform, 3
chloroplast, 9, 36
cholesterol, 29
classification, 4, 35
cloning, 20, 23
clustering, 43
coatings, 40
coding, 18
coenzyme, 15
color, iv, 13, 39
complement, 21
complexity, 29

composition, 1, 7, 9, 10, 11, 12, 13, 15, 19, 21, 22, 27, 32, 36, 37, 38, 39, 40, 43
compounds, 3, 6, 8, 10, 29, 37
condensation, 3, 9, 14
configuration, 31
consumption, 22, 24
cooking, 12, 37
copyright, iv
correlation, 26
corrosion, 40
cortex, 7
cosmetics, 8
cost, 19, 23
covering, 19
crops, 20, 22, 24, 27, 29, 31, 34, 37, 43
crude oil, 25
cues, 18
cuticle, 7, 37
cutin, vii, 7, 39
cytoplasm, 15

D

damages, iv
database, 32
defence, 8
deficiency, 14
dehiscence, 36
deposition, 7
derivatives, 3
desiccation, 3
detection, 26
developmental process, 6
diacylglycerol, 13, 15, 23, 26, 32, 33, 36, 39, 42, 43
diet, 24
diploid, 13, 29
discrimination, 19
distinctness, 19, 35
divergence, 10
diversity, 19, 31, 34
DNA, 18, 19, 26, 33, 40
docosahexaenoic acid, 23

double bonds, 10

E

eicosapentaenoic acid, 15, 33
elongation, 15, 23, 24
emerging markets, vii
encoding, 17, 18, 22, 25, 32, 36, 39, 40, 42
endosperm, 17, 32
engineering, 4, 22, 25, 32, 34, 37, 38, 40, 41, 42, 43
environmental conditions, 15
enzymes, vii, 9, 14, 15, 17, 22, 25, 26, 41
EPA, 15, 23, 24
epidermis, 7
ESI, 7, 13
essential fatty acids, vii
EST, 17, 19, 33
ester, 9
ethanol, 42
ethers, 3, 5
expressed sequence tag, 19
extraction, 12, 23

F

FAD, 9
fatty acids, vii, 8, 11, 29, 33, 34, 36, 38, 40
fertilization, 10
films, 25
fish, 23, 24, 38
fish oil, 24
flavour, 12
flax fiber, 36
flooring, vii, 29
food products, vii
fragments, 21, 40
France, 14
fruits, 9
functional analysis, 37
fungi, 24

G

gene expression, 39
gene silencing, 22, 37
gene transfer, 37
genes, 14, 15, 17, 18, 19, 20, 23, 24, 25, 26, 33, 35, 36, 42
genetic diversity, 13, 19, 31, 34
genetics, vii, 1, 32
genome, 19, 20, 21, 23, 26, 29
genomic regions, 20
genomics, vii, 1, 21, 26, 29, 38, 40, 41
genotype, 36
germination, 3, 42
glycans, 6
glycerol, 6, 13, 14, 22
glycolysis, 9
grouping, 35

H

haploid, 23
heart disease, 29
height, 7
host, 21, 24, 25, 33
hydrocarbons, 5
hydrolysis, 4
hydroperoxides, 9
hydroxyl, 25, 26
hydroxyl groups, 25

I

ideal, 19, 26, 27
India, 14
induction, 35, 39, 41
industrial chemicals, 25
infants, 24
infrared spectroscopy, 40
inheritance, 39
International Atomic Energy Agency, 39
isolation, 41
isoprene, 4
isozyme, 43

J

Jordan, 22, 37

K

ketones, 7
kinetics, 40

L

lesions, 26
lignans, vii
lipids, vii, 3, 6, 7, 8, 9, 10, 13, 29, 32, 35, 38, 42
localization, vii, 1
locus, 31
lubricants, 25

M

majority, 15
manipulation, 21, 22, 42
manufacturing, vii
mapping, 33, 39
membranes, 3
metabolic pathways, vii
metabolism, 22, 36
metabolites, 5, 6, 8, 34
molecules, 4
monolayer, 10
mRNA, 22
mutagenesis, 26, 39, 43
mutant, 13, 26, 32, 35, 36, 38
mutation, 26, 33

N

naming, 10
neutral lipids, 7, 9

next generation, 20
North America, 36
nucleus, 20
nutrition, 33, 39

O

oil, vii, 1, 7, 10, 11, 12, 13, 15, 19, 20, 21, 22, 23, 24, 25, 26, 27, 29, 32, 33, 34, 36, 37, 40, 41, 42, 43
oil samples, 7, 13
oilseed, 1, 7, 11, 12, 15, 20, 22, 24, 27, 29, 34, 37, 42
olive oil, 12
omega-3, vii
opportunities, 29
organ, 3, 4
organelles, 9
organic solvents, 3
ovaries, 15, 18
oxidation, 12

P

paints, vii, 29
parenchyma, 7
partition, 7
pathogens, 10
pathways, 23, 29, 32
permission, iv
permit, 19, 20
phenotype, 35
phenylalanine, 43
phloem, 7
phosphatidylcholine, 25
phospholipids, 9, 23, 33
phosphorylation, 41
photosynthesis, 8
physiology, 4
pistil, 9
pith, 7

plants, 6, 7, 8, 10, 14, 15, 22, 23, 24, 27, 32, 33, 34, 35, 36, 37, 38, 40, 41, 42, 43
plasma membrane, 36
point mutation, 17, 26, 42
polarization, 36
pollen, 10, 34, 37, 39
pollen tube, 10, 34
polymer, 7
polymorphism, 18, 19, 26, 43
polymorphisms, 18, 33
polypeptide, 18
polyunsaturated fat, 31, 36, 37, 38, 42, 43
polyunsaturated fatty acids, 31, 37, 42, 43
polyurethane, 40
pools, 10, 15, 26
project, 27
proliferation, 21
promoter, 14
properties, 15, 29, 40
proteins, 14, 25

R

rape, 40, 42
reactions, 14
recommendations, iv
recycling, 36
regeneration, 22
resistance, 20, 35, 41
resources, 17, 20, 23, 25, 27
restriction fragment length polymorphis, 18, 19
reticulum, vii, 9, 15, 18
rights, iv
risk assessment, 31
RNAi, 22, 27

S

saturation, 10

Index

screening, 13, 27, 38
seed, vii, 3, 7, 10, 11, 13, 15, 17, 19, 20, 22, 23, 24, 25, 27, 29, 32, 33, 34, 35, 36, 37, 39, 40, 41, 42, 43
septum, 7
sequencing, 14, 20, 21
signalling, 8
siRNA, 35
Southern blot, 14
species, 3, 7, 10, 12, 13, 17, 19, 24, 25, 33, 34, 43
stamens, 9
sterols, 4
stigma, 10
storage, vii, 3, 4, 7, 10, 13, 29, 42
strategy, 20
suberin, vii, 39
substitution, 26
substrates, 9, 15, 24
sucrose, 39
susceptibility, 20
synthesis, vii, 9, 14, 23, 24, 32, 34, 36, 37

T

tags, 32
temperature, 36, 38
testing, 41
tissue, vii, 3, 38
tobacco, 36, 43
traits, 19, 20, 27, 39
transcripts, 18
transesterification, 9
transformation, 22, 39
trends, 1, 40
triglycerides, 40
turnover, 9

V

validation, 21
variations, 10, 12
vector, 21, 40
vegetable oil, 10, 13, 32

W

wild type, 18, 42

X

xylem, 7

Y

yeast, 18, 23, 37, 41, 43